Atoms into Ions

If careless picnickers leave a sardine tin on the grass when they drive away, visitors to the same place a few weeks later will find the beauty spot spoilt by a rusty tin. In time it will fall away to a powder and cease to be an eyesore.

If a polythene bottle were left in the same place it would remain indefinitely. As you will have learnt, rusting is a property of iron but not of polythene. You will also know that the rusting of iron depends on the atmosphere as well as on the iron itself.

All this is fact which is readily established as the result of observation and experiment. This Background Book is mostly concerned with theories—theories which help to explain the rusting of iron together with a great many other phenomena. You can test facts yourself by carrying out your own experiments. You can test theories by using them to see if they help to account for the facts. Always be critical of theories, including the theories in this book. Chemists have found them useful in explaining many chemical reactions. You may also find them useful, but you must not treat them as irrefutable.

They are very important but only so far as they help us to understand what is going on — that is, the facts.

Part I
The theory of oxidation and reduction

The origin of the terms – The word 'reduction' was used in chemistry long before the word 'oxidation'. It was used to refer to the extraction of metals from their ores. In 1741 a book on assaying, the process of estimating quantitatively the amount of metal in a mineral or alloy, stated: 'Metals destroyed, and changed into Scoria or Ashes, are by their union with the same matter again restored to their metallick form. This operation is called reduction.'

A sixteenth-century assayer's touchstone and needles. The touchstone is still used by gold-smiths today to test the purity of gold.

The gold is rubbed on the stone and the colour of the streak left behind indicates the amount of base metal alloyed with the gold. *Science Museum*

Before the discovery of oxygen the reverse process of the conversion of the metal to its oxide was called 'calcination,' the oxide being called the 'calx'.

Following the discovery in 1777 of the true nature of combustion, and the identification of the new element oxygen, the word 'oxidation' was used by Lavoisier to describe the combination of substances with oxygen.

By the beginning of the nineteenth century, therefore, the terms 'oxidation' and 'reduction' were being used to describe the process by which elements or compounds combined with the element oxygen to form oxides, and the reverse of this process, by which the element oxygen was removed from a compound. At this stage it is worth considering what substances are likely to remove oxygen from the oxide and reduce it. The elements carbon and hydrogen immediately spring to mind as they are substances which have a great affinity for oxygen. Thus, copper oxide is readily reduced to copper by heating either in a stream of hydrogen or mixed with charcoal.

$$CuO(c) + H_2(g) \rightarrow Cu(c) + H_2O(g)$$

$$2CuO(c) + C(c) \rightarrow 2Cu(c) + CO_2(g)$$

Such substances, which bring about the reduction of oxides, are called 'reducing agents'. You should be able to make a list of several. You will notice in the above reactions that when the hydrogen and the carbon reduced the copper oxide to copper they were themselves oxidized to water and carbon dioxide. The total reaction was both a reduction and an oxidation. You will see later that in all these reactions oxidation must accompany reduction; as a result they are often called *redox* reactions. Another point, clearly brought out in the reduction of

The last stage is electrolytic refining, which produces 98·98 per cent pure copper; thin copper cathodes are inserted between anodes of impure copper in copper sulphate solution. During electrolysis copper is transferred from anode to cathode while impurities remain in solution or collect as a sludge beneath the anode.
Copper Development Association.

Refining copper.
Oxygen is removed in the final stages of refining by 'poling'; tree trunks are consumed in the molten copper, reducing copper oxide to copper.
Copper Development Association.

In modern steel manufacture, iron is reduced from its ore in the presence of coke and limestone in a blast furnace.
British Iron & Steel Federation

copper oxide, is that the hydrogen and carbon were oxidized not by oxygen gas but by oxygen from a chemical compound. Any substance which in this way supplies oxygen for the oxidation process is called an 'oxidizing agent'. You should be able to make a list of oxidizing agents.

Redox reactions have many features in common. For example they are nearly always accompanied by the liberation of heat. Make a list of all the sources of energy you can think of and see how many of them involve a redox reaction. The pictures on the page opposite show one redox reaction where much energy is liberated. The photograph on page 17 shows another.

Since the beginning of the nineteenth century it has been convenient to include many more chemical changes under the general title of redox reactions. When magnesium is heated in chlorine it catches fire, burning much as it does in air and forming a white powder.

$$Mg(c) + Cl_2(g) \rightarrow MgCl_2(c)$$

When magnesium is heated in nitrogen it forms a yellow powder of magnesium nitride. Similarly mixtures of magnesium powder and sulphur are dangerous, burning explosively when ignited to form magnesium sulphide, which is a white powder.

$$3Mg(c) + N_2(g) \rightarrow Mg_3N_2(c)$$

$$Mg(c) + S(c) \rightarrow MgS(c)$$

In these three reactions, as in the reaction with oxygen, the magnesium has combined with a non-metallic element, heat and light have been emitted, and a powder has been formed. Because they have so many points in common it is convenient to consider them all under the heading of oxidation.

The electronic interpretation of redox reactions – The interpretation of these changes can be given in terms of the atomic theory of matter and the bonding of atoms. The structures of magnesium oxide, magnesium chloride, magnesium nitride, and magnesium sulphide are all ionic; they contain the ion Mg^{2+} and a negative ion, O^{2-}, S^{2-}, N^{3-}, or Cl^-. Hence in each of the reactions each magnesium atom has lost two electrons to become a magnesium ion. At the same time non-

Thermit reactions involve the reduction of metallic oxides by the use of finely divided aluminium powder. In the process the aluminium is oxidized and the metallic oxide is reduced. This process is used on railways to weld rails together.

A mixture of iron oxide and aluminium powder is placed between the rails, and a small amount of a mixture of sodium peroxide and aluminium powder is placed above this.

The mixture is ignited and the aluminium is almost instantaneously oxidized with the evolution of a great deal of heat. The iron produced is in a molten state and on cooling welds the rails together.

British Rail

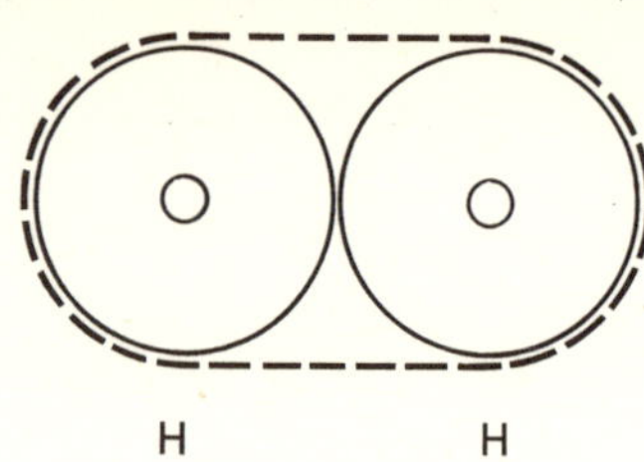

The H$_2$ molecule; the electron cloud is equally shared by both hydrogen atoms.

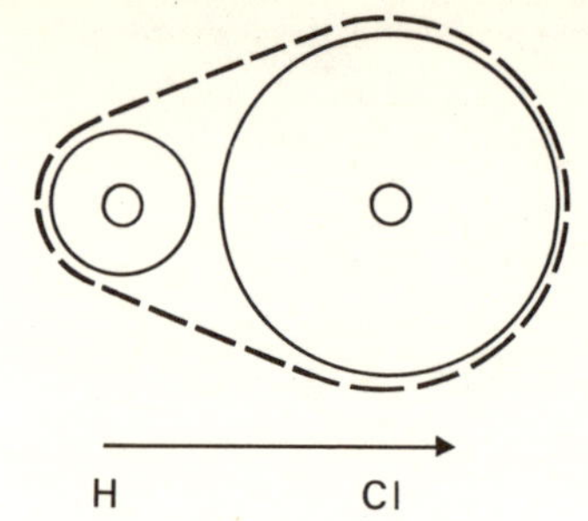

The HCl molecule; the electron cloud is attracted towards the chlorine atom.

metallic atoms have gained electrons to become negative ions:

$$Mg \rightarrow Mg^{2+} + 2e^-$$

$$O_2 + 4e^- \rightarrow 2O^{2-}$$

$$MnO_4^- + 8H^+ \rightarrow Mn^{2+} + 4H_2O + 5e^-$$

$$Cl_2 + 2e^- \rightarrow 2Cl^-$$

$$N_2 + 6e^- \rightarrow 2N^{3-}$$

$$S + 2e^- \rightarrow S^{2-}$$

These are called 'half equations' as they only show the change taking place with one of the two reagents. Two half equations can be combined to give the overall equation. Since each equation involves a transfer of electrons from the metal to the non-metal, an overall equation can be formulated by adding together two half equations in such a way that there is no surplus or deficiency of electrons. To make an overall equation out of the first two equations above, the first equation must be multipled by 2.

$$2Mg \rightarrow 2Mg^{2+} + 4e^- \qquad \text{half equation}$$

$$O_2 + 4e^- \rightarrow 2O^{2-} \qquad \text{half equation}$$

$$O_2 + 2Mg \rightarrow 2Mg^{2+} O^{2-} \qquad \text{overall equation}$$

Since magnesium loses electrons when it combines with oxygen, we might generalize by saying all oxidation reactions involve the loss of electrons.

The extraction of magnesium metal from any of the four compounds is a reduction process which involves the removal of a non-metallic element. In this case the magnesium ions will have regained their electrons:

$$Mg^{2+} + 2e^- \rightarrow Mg$$

To generalize we might say that according to the atomic theory, a substance is reduced when it has gained electrons. It follows that in the first four reactions the non-metal has been reduced to its ions by gaining electrons. As electrons are only transferred from one chemical substance to another, one substance must gain electrons as the other loses them: it is impossible to have oxidation without reduction. This is the conclusion we reached when considering oxidation and reduction in terms of gain or loss of oxygen.

Oxidation is a Loss of electrons
Reduction is a Gain of electrons

For this way of looking at the changes to be really useful it must be extended to include substances which are covalently bonded as well as those which are ionic. A covalent bond involves the sharing of a pair of electrons between two atoms. If the two atoms are of the same element, such as the diatomic molecules of hydrogen or chlorine, the electrons are shared equally between two atoms. In all other cases they are not shared equally; the electrons are to be found slightly closer to one atom than to the other. Because this atom seems to have the greater affinity for electrons, and because electrons are defined as negative electrical quantities, the electrons in this case are said to be closer to the more 'electronegative' of the two atoms, A molecule composed of these atoms is said to be 'polarized'. Chlorine atoms have a greater attraction for electrons than have hydrogen atoms. In the hydrogen chloride molecule, the bond is polarized in such a way that the electrons are slightly closer to the chlorine than to the hydrogen.

The electrochemical series is a series of the elements in the order of the tendency of atoms to lose electrons and form positive ions. If this tendency is great, the element is said to be 'electropositive'. Thus, the metals and hydrogen are electropositive elements, since they tend to form positive ions. Similarly, an element is said to be 'electronegative' if its atoms tend to gain electrons and form negative ions. Many non-metals are electronegative. In a covalent compound of two elements, the more electronegative element gets a greater share of the electrons than the more electropositive, e.g.

small share of one electron   H—Cl   large share of one electron

Oxygen has a covalency of two so that in covalent compounds with more electropositive elements it gets the largest share of two electrons.

large share of two electrons   O   small share of each electron

Redox reactions involving covalent molecules can now be interpreted in terms of the transfer of electrons. The burning of methane to form carbon dioxide serves as a good example:

$$CH_4 + 2O_2 \rightarrow CO_2 + 2H_2O$$

H
|
H—C—H
|
H

Carbon is more electronegative than hydrogen, so has the major share of four extra electrons.

O—C—O

Carbon is less electronegative than oxygen, so loses some of its share of four electrons.

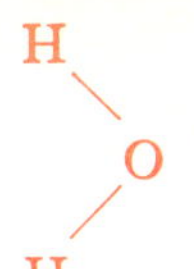

Each hydrogen atom loses some of its share of one electron whilst the oxygen gets a larger share of two electrons.

Hence the carbon in the methane has been oxidized because it has changed from the state of having a surplus of four electrons to one of having a deficiency of four electrons in carbon dioxide. The oxygen has been reduced from the state of no extra electrons in the element itself to two extra electrons in both the water and the carbon dioxide. The hydrogen in the methane has been neither oxidized nor reduced because both in methane and in water it has a small share of one electron.

Corrosion—The existence of life on this planet is almost totally dependent on our oxidizing atmosphere. It is this same oxidizing atmosphere which has the very great disadvantage of causing the corrosion of metals.

All corrosion of metals involves the oxidation of the metal to its ions.

$$Fe \rightarrow Fe^{2+} + 2e^-$$

Subsequently may things may happen to the ions; in the case of iron the ferrous ions are eventually converted to hydrated ferric oxide, which is rust. The oxidation is brought about by the oxygen in the atmosphere, and of course the oxidation of the metals is dependent on their ability to lose electrons.

Enormous advances have been made in treating metals in such a way that they are less liable to corrosion, but it is only too obvious when one looks at an old car or bicycle that the problem is by no means solved. Possibly in some spheres the ultimate solution will lie in finding materials, perhaps plastics, which have the structural strength of metals but which are not readily able to lose electrons to the oxygen in the atmosphere.

Part II
Oxidizing and reducing strength

In certain circumstances manganese(II) ions, Mn^{2+}, can be oxidized to permanganate ions, MnO_4^-, and under different conditions permanganate ions can be reduced back to manganese(II) ions. We also know that chloride ions can be oxidized to chlorine and vice versa.

$$Mn^{2+} + 4H_2O \rightarrow MnO_4^- + 8H^+ + 5e^-$$

$$2Cl^- \rightarrow Cl_2 + 2e^-$$

But there is nothing here which will tell us whether chlorine will oxidize manganese(II) ions or permanganate ions oxidize chloride ions. The answer to this sort of problem requires consideration of electrolytic processes.

In electrolysis and in electrolytic cells the reactions taking place at the electrodes involve the gain or loss of electrons. In an electrolytic cell the negative electrode is the one at which electrons leave the cell: the process at the negative electrode involves the liberation of electrons and hence, by definition, is an oxidation. Similarly the process at the positive electrode,

A simple copper-zinc cell.

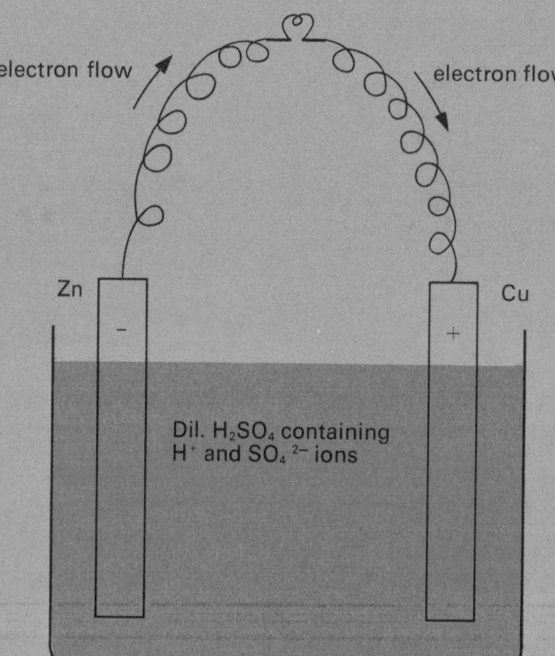

which involves the taking up of electrons, is by definition reduction. The simple cell of zinc and copper dipping into
dilute sulphuric acid illustrates this clearly. The zinc electrode
is negative. In effect electrons have been transferred from zinc
metal to hydrogen ions.

$$Zn + 2H^+ \rightarrow Zn^{2+} + H_2$$

This is a process which one might expect to be reversible. The
fact that it takes place in the direction given suggests that zinc
has a greater tendency to lose electrons and form its ions in
solution than has hydrogen. The electromotive force of this
cell, measured in volts, is found to be a little greater than one
formed by dipping iron and copper into dilute sulphuric acid.
As the process taking place at the positive electrode is the
same in the two cases, the difference in e.m.f. must depend on
the difference in the processes taking place at the negative
electrodes. This implies that zinc has a greater tendency to
form its ions in solution than has iron. You will almost cer-
tainly have met this idea before when studying electrolysis
and you will probably have expressed it in the words 'zinc has
a higher electrode potential than iron', or 'zinc is higher in the
electrochemical series than iron'. What is now apparent is that
this is the same as saying that zinc is more readily oxidized to
its ions than iron. It is possible to make any redox process the
electrode process of an electrolytic cell. The strength of this
as an oxidation process can therefore be given in terms of an
electrical potential. We use the expression 'oxidation potential'. This term covers all the electrode potentials you have
met previously. It is not essential to have a precise definition
of 'oxidation potential' at this stage; it is sufficient to know
that the greater the positive electrical potential, the stronger
it is as an oxidation process. Equally, the greater the negative
potential, the stronger the reverse reaction as a reduction process.

A list of oxidation potentials is given in the table on page 16.
Before we make use of the table it must be emphasized that the
e.m.f. of a cell depends on the concentration of ions in the
electrolyte and on the temperature, and oxidation potential
is defined under the precise conditions of these concentrations
and temperatures. These may not be the conditions under
which we want to consider a particular reaction. In general,
however, if the oxidation potential of process A is greater than

process B there should be conditions of concentration and
temperature under which process A will cause process B to
proceed in the reverse direction.

We can now answer the question we posed at the beginning
of this section: will permanganate ions oxidize chloride ions?
The table gives us the following oxidation potentials:

$$MnO_4{}^- + 8H^+ \rightarrow Mn^{2+} + 4H_2O + 5e^- \qquad +1{\cdot}52 \text{ volts}$$

$$Cl_2 + 2e^- \rightarrow 2Cl^- \qquad +1{\cdot}36 \text{ volts}$$

We would therefore predict that the answer is 'Yes' and this
is what we find in practice. A simple source of chlorine in the
laboratory is the action of concentrated hydrochloric acid on
crystals of potassium permanganate.

This table is of value in interpreting many reactions and
you should practice using it. Here are two examples.

1. Which of the following could be used to oxidize iron(II)
ions to iron(III) ions? a. sulphur, b chlorine, c. bromine,
d. iodine, e. potassium dichromate.

2. Which of the following gases will turn an aqueous solution
of potassium dichromate green (i.e. reduce it to Cr^{3+})? a.
H_2S, b. HCl, c. HI, d. SO_2, e. HF, f. HBr (In solution halogen
acids will ionize to halide ions.)

Part III
The use of electrolytic oxidation and reduction

In the previous section it was shown that in electrical cells the processes at the positive electrode were reductions and at the negative electrode were oxidations. In the case of electrolysis the electrons enter the cell through the negative electrode which is called the 'cathode'; and they leave the cell at the positive electrode, which is called the 'anode'. Thus the cathode reaction is a reduction and the anode reaction is an oxidation. Useful reduction and oxidation reactions may be brought about by electrolysis.

The production and plating of metals – The liberation of a metal from its ores must involve a reduction process. Electrolysis can be a more powerful tool for reduction than any other process and it is for this reason that all the highly reactive metals are produced by electrolytic processes. Potassium, sodium, calcium, magnesium, and aluminium are manufactured by the electrolysis of a suitable molten salt. If the conditions of electrolysis are carefully controlled, it may be possible to deposit the metal on the cathode as a smooth coherent layer; this is the method used to plate the surface of one metal with another. It is known as 'electrodeposition'. Copper, nickel, and chromium are all plated by this method.

The production of chlorine and fluorine – The conversion of sodium chloride to chlorine involves an oxidation.

$$2Cl^- \rightarrow Cl_2 + 2e^-$$

Most chlorine is manufactured by the electrolysis of chlorides, the chlorine being liberated at the anode.

You will see that fluorine is at the very bottom of the table of oxidation potentials on page 16. This means that the only way of liberating fluorine from fluorides is by electrolytic oxidation. In the manufacture of fluorine it is evolved from the anode when a mixture of potassium and hydrogen fluorides are electrolysed. Great difficulty is experienced in the process, as fluorine is such a strong oxidizing agent that it tends to oxidize anything with which it comes into contact. It has been found that copper is only attacked on the surface, forming a layer of fluoride which protects the rest of the metal, so the electrolysis vessel may be constructed of copper.

Anodizing aluminium – If aluminium is made the anode in the electrolysis of a solution of chromic acid, or sulphuric acid, it is oxidized on the surface to a depth of only a fraction of a millimetre. This layer of oxide is so thin that it is hardly apparent to the naked eye. Aluminium treated in this way is more resistant to corrosion than untreated aluminium. Moreover it will take a dye; highly coloured aluminium trays, or saucepan lids, are often made from anodized aluminium. The general properties of the aluminium are not altered by anodizing.

The production of hydrogen peroxide – Hydrogen peroxide is a

"

Aluminium is anodized in tanks
such as the one shown in the
centre, from which an aluminium
sheet is just being removed.
Acorn Anodizing Company Ltd.

strong oxidizing agent close to the bottom of the table of oxidation potentials.

$$H_2O_2 + 2e^- + 2H^+ \rightarrow 2H_2O$$

It is now used a great deal as the oxidant in rocket fuels where it is required in an undiluted form. In World War II the Germans found a way of manufacturing it in a pure enough form.

At first, attempts to concentrate solutions of hydrogen peroxide failed because it started to decompose to water and oxygen.

$$2H_2O_2 \rightarrow 2H_2O + O_2$$

This reaction is catalysed particularly by the presence of metal ions, and the methods of manufacture left foreign ions in the solution.

An early method of overcoming this involved the electrolysis of sulphuric acid of fairly high concentration – 50 to 60 per cent. Under these conditions the bisulphate ion is discharged at the anode giving persulphuric acid.

$$2H_2SO_4 \rightarrow 2H^+ + 2HSO_4^-$$

$$2HSO_4^- \rightarrow H_2S_2O_8 + 2e^-$$

The product readily reacts (on warming) with water to give hydrogen peroxide and sulphuric acid.

$$H_2S_2O_8 + 2H_2O \rightarrow H_2O_2 + 2H_2SO_4$$

After the electrolysis the anode solutions are distilled at a reduced pressure and a solution of hydrogen peroxide in water is collected. It is very pure and the hydrogen peroxide can now be concentrated by further distillation. The overall equation obtained by adding the three equations together is:

$$2H_2O \rightarrow H_2O_2 + 2H^+ + 2e^-$$

Water has in fact been oxidized to the peroxide and the oxidation stage was at the anode during the electrolysis.

Fuels and sources of energy — When we think of fuels we tend to have in our minds only substances such as coal and gas which can be burnt to provide heat. Fuels, of course, supply most of man's energy and include the food he eats to provide the energy he needs for movement, the materials he burns to keep himself warm, and the materials he uses to provide the motive power for all the machinery employed to make his day's work easier. There are only a few sources of energy, such as hydroelectric energy, wind and tidal energy, and atomic energy, which do not depend on fuels and even here we sometimes talk about uranium 235 or plutonium as an atomic fuel.

When fuels liberate energy they do so by undergoing chemical change and nearly every such change is a redox reaction.

PARSONS

It is the earth's atmosphere which really makes all this possible, as all conventional fuels depend on atmospheric oxygen as the oxidizing agent. By far the most common fuels are the organic compounds containing carbon and hydrogen and we know them mostly in the form of wood, coal, and oil; throughout the history of mankind they have been burnt to liberate the energy of the reaction in the form of heat. It was only with the coming of the steam engine and then the internal combustion engine that it became possible to harness the heat energy and make use of it either directly for driving machinery or indirectly by generating electricity. It is interesting to trace transfers of energy in a coal-fired power station. The coal is burnt in furnaces (the redox part of the process), the heat from the furnaces raises steam in boilers, and the energy of the steam drives the turbines. The turbines provide the motive power for dynamos which finally convert the energy into electricity.

This mechanism for generating electricity requires a great deal of very complicated machinery which must be precision-built. Yet it all depends on the liberation of energy by the transfer of electrons from one substance to another in the burning of the coal. If methane were used as the fuel the half equations would be:

$$CH_4 + 2H_2O \rightarrow CO_2 + 8H^+ + 8e^-$$

$$O_2 + 4H^+ + 4e^- \rightarrow 2H_2O$$

Recently it has been found possible to use reactions like these as the anode and cathode reactions of an electrolytic cell, so that instead of the energy being liberated as heat it is converted directly into a flow of electrons. Obviously the simple cell shown on page 8 does just this, but the fuel is a metal, and a metal is much too expensive to make an economical fuel. A fuel cell can be defined as one in which conventional fuels are used to generate and use up electrons for the electrode reactions. Most fuel cells so far have used hydrogen for the negative electrode and oxygen for the positive.

$$H_2 \rightarrow 2H^+ + 2e^-$$

$$O_2 + 4e^- \rightarrow 2O^{2-}$$

The difficulty is to get hydrogen to liberate electrons steadily and oxygen to take up electrons steadily. The reactions only take place on the surface of active catalysts. In 1964 Shell Research Limited announced the development of a successful cell using methyl alcohol as the fuel. However, the methyl alcohol is not directly oxidized at the negative electrode. First it is reacted with steam in the presence of a catalyst to generate hydrogen.

$$CH_3OH + H_2O \rightarrow CO_2 + 3H_2$$

The hydrogen is supplied to one electrode and air is pumped to the other. The electrolyte is dilute sulphuric acid. The experimental unit generates a power of 5 kilowatts but at present it is only a little more efficient than a diesel electric motor in converting chemical energy into electrical energy. This is largely because of the need to first convert the methyl alcohol into hydrogen. Research is continuing to see if the methyl alcohol can be used directly.

Such methods of utilizing fuel energy would seem to have a great future, for the machinery involved is obviously fairly

The fuel cell shown here runs on methanol: it generates 5 kilowatts of electricity.
Shell.

simple compared with that involved in present day generators.

Fuel cells are perhaps the clearest evidence that in redox processes electrons are simply transferred from reducer to oxidant.

Life itself is a far more efficient mechanism for harnessing fuel energy than any man-made machine. The fuel is the food we eat, which is broken down in the digestive system to glucose, $C_6H_{12}O_6$, to be transferred to all the muscles by the blood stream. Glucose is a reducing agent:

$$C_6H_{12}O_6 + 6H_2O \rightarrow 6CO_2 + 24H^+ + 24e^-$$

It is oxidized in the muscles by oxygen which is absorbed into the blood stream in the lungs.

$$O_2 + 4H^+ + 4e^- \rightarrow 2H_2O$$

The overall reaction liberates energy which is used in the working of the muscle. The energy of all life, both plant and animal, is obtained by redox reactions of a similar type.

Oxidation Potentials

The lower in this list the stronger the oxidizing agent.

Oxidizing Agent		Reducing Agent	E°
K^+	$+e^-$	$\rightleftharpoons K$	-2.93
Ca^{2+}	$+2e^-$	$\rightleftharpoons Ca$	-2.87
Na^+	$+e^-$	$\rightleftharpoons Na$	-2.71
Mg^{2+}	$+2e^-$	$\rightleftharpoons Mg$	-2.37
Al^{2+}	$+3e^-$	$\rightleftharpoons Al$	-1.66
Mn^{2+}	$+2e^-$	$\rightleftharpoons Mn$	-1.18
Zn^{2+}	$+2e^-$	$\rightleftharpoons Zn$	-0.76
Fe^{2+}	$+2e^-$	$\rightleftharpoons Fe$	-0.44
Pb^{2+}	$+2e^-$	$\rightleftharpoons Pb$	-0.13
$2H^+$	$+2e^-$	$\rightleftharpoons H_2$	0.00
S	$+2H^+ + 2e^-$	$\rightleftharpoons H_2S$	0.14
SO_4^{2-}	$+4H^+ + 2e^-$	$\rightleftharpoons SO_2 + 2H_2O$	0.17
Cu^{2+}	$+2e^-$	$\rightleftharpoons Cu$	0.34
I_2	$+2e^-$	$\rightleftharpoons 2I^-$	0.54
MnO_4^-	$+e^-$	$\rightleftharpoons MnO_4^{2-}$	0.56
Fe^{3+}	$+e^-$	$\rightleftharpoons Fe^{2+}$	0.77
Ag^+	$+e^-$	$\rightleftharpoons Ag$	0.80
Br_2	$+2e^-$	$\rightleftharpoons 2Br^-$	1.06
MnO_2	$+4H^+ + 2e^-$	$\rightleftharpoons Mn^{2+} + 2H_2O$	1.23
$Cr_2O_7^{2-}$	$+14H^+ + 6e^-$	$\rightleftharpoons 2Cr^{3+} + 7H_2O$	1.33
Cl_2	$+2e^-$	$\rightleftharpoons 2Cl^-$	1.36
MnO_4^-	$+8H^+ + 5e^-$	$\rightleftharpoons Mn^{2+} + 4H_2O$	1.51
MnO_4^-	$+2H_2O + 3e^-$	$\rightleftharpoons MnO_2 + 4OH^-$	1.69
H_2O_2	$+2H^+ + 2e^-$	$\rightleftharpoons 2H_4O$	1.77
F_2	$+2e^-$	$\rightleftharpoons 2F^-$	2.85

The higher in this list the stronger the reducing agent.

Rockets are propelled by the oxidation of their fuels.
U.S.I.S.